KRIS NOLAN

Rocket Engine Cycles for Beginners

A Friendly Guide to Rocket Propulsion and Cycle Principles

Copyright © 2024 by Kris Nolan

All rights reserved. No part of this publication may be reproduced, stored or transmitted in any form or by any means, electronic, mechanical, photocopying, recording, scanning, or otherwise without written permission from the publisher. It is illegal to copy this book, post it to a website, or distribute it by any other means without permission.

First edition

This book was professionally typeset on Reedsy.
Find out more at reedsy.com

Contents

1 Overview	1
An Overview of Rocket Engine History	2
Fundamentals of Rocket Engines	3
Pump-Fed Cycles	4
2 Propellant & Engine Components	6
Rocket Engine Components	6
Rocket Propellant	9
How Engine Design Is Affected by Propeller Selection	11
3 Engines Cycles	13
Overview of Engine Cycles	14
A Synopsis of Cold Gas and Pressure-Fed Engines	14
Pump-Fed Cycle Engines	15
Tap-Off Cycle	26
Electric Pump-Fed Cycle	28
In brief	28
4 Real-World Applications	30
Early Day (Cold War Era)	30
Contemporary Uses	32
Upcoming Patterns	33
5 Conclusion	36
6 Reference	40

1

Overview

Greetings from Rocket Engine Cycles for Novices! A fascinating exploration of engine cycles and rocket propulsion is about to begin. If you're here, you're likely wondering how rockets operate, what drives them, and why their construction is the way it is. Don't worry; you can grasp these ideas without a Ph.D. This book is your guide, simplifying difficult concepts into understandable, approachable insights.

Rockets are incredibly powerful devices designed to defy gravity, carry payloads into space, and occasionally even transport people to far-off places. These machines' engines are their strong points, and knowing how they operate can help you appreciate the wonders of space travel.

To let you know how it all started and where we're going, let's start with some history.

An Overview of Rocket Engine History

The use of fireworks in ancient Chinese festivals and celebrations can trace the origins of rockets back thousands of years. However, it was not until the 20th century that rocketry started to develop as a science, owing to the efforts of trailblazers such as Robert Goddard. Goddard's research on liquid-propellant rockets laid the foundation for contemporary rocketry in the 1920s. Picture a fuel-filled tube launching into the air. Goddard's initial rockets were straightforward but groundbreaking.

During World War II, Germany created the first long-range guided ballistic missile, the V-2, which genuinely drew the world's attention to rockets. The V-2 demonstrated that rockets might travel to new heights and, eventually, beyond our atmosphere when they reached the edge of space before descending back to Earth.

The space race of the 1950s and 1960s followed. There was intense competition between the US and the USSR to send probes, launch satellites, and finally land humans on the moon. Rockets power all these landmarks. Rockets are still necessary to reach and comprehend the last frontier, even if private organizations like SpaceX and Blue Origin join NASA in exploring Mars and beyond.

OVERVIEW

Fundamentals of Rocket Engines

A rocket engine's primary function is propulsion. Consider inflating a balloon and releasing it. The balloon takes off in the opposite direction as the air pours out one end. The engine nozzle releases hot gas, creating an opposing force that propels the rocket forward. This is comparable to how a rocket engine operates. We can summarize Newton's Third Law of Motion as follows: "Every action has an equal and opposite reaction."

Rocket engines burn fuel and an oxidizer to produce the swift stream of gas that propels the rocket in the opposite direction. For a general idea, consider this simplified equation:

Thrust (F) = Mass of propellant (m) x Acceleration (a)

Consider the propellant to be the engine's lifeblood or fuel. The engine ignites this fuel, producing expanding gas that swiftly escapes through the nozzle, generating thrust.

There are too many different rocket engine cycle variations available in the rocket industry. Engine cycles range from pump-fed to pressure-fed from cold gas thrusters. However, this book will never cover cold gas thrusters and pressure-fed engine cycles. One particular type of engine cycle is far more dependable to use when it comes to cargo capacity and orbital reach.

To be more specific, two primary categories of engines use propellant: pressure-fed engines, which use fuel tanks to supply propellant to the combustion chamber, and pump-fed engines, which use pumps and turbines to increase flow rate and efficiency. This book primarily focuses on pump-fed engines, as they facilitate the creation of larger, more powerful designs commonly seen in modern rockets.

Pump-Fed Cycles

There are various ways to feed fuel to a rocket engine, and understanding these techniques reveals the benefits of each type. The most basic engines, such as cold gas thrusters, simply release gas from a pressurized tank without the need for burning. Others, such as pressure-fed engines, force propellant straight into the combustion chamber using high pressure. However, the pump-fed engines are more sophisticated and effective.

A pump-fed engine generates significantly more power by moving the propellant from the tanks into the combustion chamber via a pump. For missions requiring long-distance, high-speed propulsion, such as delivering cargo into orbit or beyond, this is essential. We will mostly concentrate on pump-fed cycles since they are the foundation of the majority of contemporary rockets, but you will also get a taste of cold gas and pressure-fed engines. You'll see why pump-fed cycles are the preferred option for high-stakes space travel once you comprehend them.

When I first began learning about rocket engines, the technical jargon and their overwhelming complexity intimidated me. For novices, rocketry can feel like entering a foreign land because it has its own language. I wrote this book to introduce anyone interested in science, engineering, or space to the world of rocket engine cycles.

My objective is to walk you through these concepts in an easy-to-understand and entertaining manner, regardless of whether you're a high school student, a hobbyist, or just interested in rockets. You will have a firm understanding of rocket engine cycles and how each part functions in tandem to enable a rocket to take off. Are you ready? Let's begin!

2

Propellant & Engine Components

Despite their high-tech appearance, rockets consist of various parts, each serving a distinct purpose. Every component of a rocket engine, from the fuel tanks to the nozzle, helps produce and direct thrust. This chapter will cover every major part of a rocket engine, providing an explanation of its function and significance. Next, we will delve deeper into propellants, and the fuel that powers these devices, and examine how the propellant selection influences the overall design of an engine.

Rocket Engine Components

A rocket engine is essentially a collection of components that cooperate to produce strong thrust from stored fuel. Let's examine each essential element.

Propellant Tank

The propellant tank stores the fuel and oxidizer, which are essentially the "food" that the engine needs to run. A rocket needs a place to store its propellant, much like a car needs a petrol tank. Rocket propellant tanks, on the other hand, are much more specialized. Often designed to withstand high temperatures and pressures, they are lightweight to avoid adding unnecessary weight to the rocket. Fuel (like liquid hydrogen or kerosene) and an oxidizer (like liquid oxygen), typically stored in a tank, combine to produce the rocket's explosive force. We carefully select the materials and insulation of these tanks to withstand the severe conditions of space, ensure strong pressure resistance, and prevent fuel leaks.

Pump

Pumps move the propellant from the tanks into the combustion chamber. Consider the pump to be the rocket engine's core component. The pump transports the propellant from the tanks to the combustion chamber, just like your heart does while pumping blood throughout your body. Rocket pumps are strong. To maintain a steady and regulated flow, they must push the fuel and oxidizer at very high pressures. The creation of the powerful thrust required to launch a rocket depends on this pressurized flow.

Combustion Chamber

The "magic" that creates propulsion is the combustion chamber, where oxidizer and fuel combine and ignite. The combustion chamber functions at temperatures hot enough to melt steel, much like an oven. The propellant combines and ignites at this point. High-speed gas, released via the nozzle when combustion begins, produces the thrust of the rocket. Engineers design these chambers to withstand high pressure and temperatures. To keep the chamber from melting, engineers employ high-strength materials, frequently in conjunction with a cooling system.

Injector

For the best burn, the injector's job is to evenly spray the fuel and oxidizer into the combustion chamber. Consider a showerhead with small nozzles that spray water. Similar to this, the injector releases a fine mist of oxidizer and fuel to promote even mixing and effective combustion. A well-designed injector ensures consistent and uniform mixing of the fuel and oxidizer. Uneven fuel can result in incomplete combustion, energy waste, or equal hazardous engine problems; therefore, this mixing is essential.

Igniter

The igniter initiates the entire thrust process by igniting the initial combustion. The engine would not start if there was no igniter. The igniter, a compact yet potent mechanism, provides an initial spark or heat that ignites the fuel and oxidizer mixture and initiates combustion. Consider it analogous to an automobile engine's spark plug. Once lit, the combustion process becomes self-sustaining as long as the fuel and oxidizer continue to flow.

Nozzle

The nozzle directs the gas produced by the combustion chamber to maximize thrust. In order to expand and accelerate the hot gas and provide them with a directed path out of the engine, the nozzle is frequently bell-shaped. This concentrated jet of gas out of the nozzle propels the rocket forward. In order to maximize thrust efficiency and stability, engineers carefully design and structure the nozzle to manage how gas expands.

Cooling System

By controlling the high temperatures produced during combustion, the cooling system prevents the engine from overheating. Rocket engines would rapidly overheat without a cooling system because they run at thousands of degrees. Certain engines feature cooling channels incorporated into the combustion chamber walls, which let fuel pass through and absorb heat. It's similar to putting cold water through a coffee maker to keep it from getting too hot. The engine's performance and stability depend on this cooling procedure.

Rocket Propellant

After examining the parts, let's examine the rocket's fuel source. Every engine cycle requires propellants, and each type has advantages and disadvantages. Here, we'll examine the primary varieties, their distinctive characteristics, and their effects on engine design.

Propellant Types

Solid, liquid, or hybrid propellants are the three types commonly used in rocket engines.

Solid Propellant: stable mixtures of oxidizer and fuel. Booster stages or smaller rockets frequently use them. Once lit, they continue to burn until the fuel runs out, despite their reliability, simplicity, and convenience in storage. Notable examples are Zinc-Sulphur (ZS), Black Powder (Gunpowder), Composite Modified Double-Base (CMDB), Composite Propellant (AP & HTPB mix), and Double-Base Propellant (NG & NC mix).

Liquid Propellant: Many large rockets (like the Space Shuttle) use liquid propellants, such as liquid hydrogen and liquid oxygen. These propellants are perfect for complex missions since they provide more control because they may be turned on or off as needed. However, its design is complicated by the need for high-pressure pumps and cryogenic storage. Notable examples are RP-1, LCH4, cryogenic propellants (LOX + LH2), hypergolic propellants (N2O4 + MMH & N2O4 + UDMH), and storable liquid propellants (IRFNA + UDMH).

Hybrid Propellant: Solid fuel and liquid or gaseous oxidizer are both used in hybrid rockets. They are a flexible option since they provide part of the control of liquid rockets and some of the simplicity of solid rockets.

The propellant's characteristics

Performance is influenced by the propellant type's energy density, or how much energy it can store, and burn rate, or how quickly it can release energy. These characteristics affect rocket performance in the following ways:

Greater **energy density** allows for the use of less fuel to produce more power. For instance, liquid hydrogen is a common option for long-distance missions due to its high energy density. But because of its low density, hydrogen likewise needs big, well-insulated containers. **Burn Rate:** The propellant's burn rate determines its rate of combustion. While a slower rate can enable longer, sustained burning, a quicker rate can instantly produce tremendous thrust. Because it directly affects thrust and efficiency, engineers take the burn rate into account while building the engine cycle.

How Engine Design Is Affected by Propeller Selection

The specifications for various launch missions vary, and the engine design may alter according to the propellant selection. Each type of propellant influences the rocket's architecture differently. Here's how:

Solid Propellants: The inability to shut off in midair makes solid propellant rockets less flexible, but also simpler. They are lighter and more dependable for some tasks, like launching small payloads, because they don't need pumps or intricate fuel connections.

Liquid propellants add complexity but provide you control over the burn because they need pumps, tanks, and precise injectors. For example, because hydrogen occupies more space, liquid hydrogen engines have larger tanks, whereas liquid oxygen requires insulation because of its cryogenic temperature. When multiple stages are needed or for human missions, liquid propellants provide the necessary precision.

Hybrid Propellants: These provide adaptability by having a solid fuel core and a liquid or gas oxidizer. They are appropriate for some commercial and research applications because they require more moving components than solid engines but fewer than liquid engines. Selecting the best propellant involves weighing mission requirements, cost, efficiency, and safety in addition to achieving the highest thrust.

This chapter covered the essential parts of a rocket engine and the range of propellants that power them. With this basic understanding, you're prepared to delve further into the intriguing realm of rocket engine cycles, where every decision regarding fuel, design, and cycle type significantly affects a rocket's range and speed.

3

Engines Cycles

Welcome to engine cycles, the core component of rocket engine design. An engine cycle explains how a rocket engine controls the combustion gas generated after injecting propellant into the combustion chamber. Put more simply, it's the "plan" or "strategy" that a rocket engine employs to produce thrust that is dependable, strong, and efficient.

Every cycle type has a specific set of objectives. While some are sophisticated yet incredibly effective, others are more straightforward and dependable. This chapter will cover pump-fed cycles, the true powerhouses of contemporary rocketry, after we've covered some basic cycle types, such as cold gas and pressure-fed engines.

Overview of Engine Cycles

Engineers use engine cycles to determine the optimal way to manage the fuel, combustion, and exhaust processes to satisfy the particular needs of every mission. The engine cycle controls the rocket's fuel efficiency and thrust production, regardless of whether it is launching a probe far into space or a modest cargo into low Earth orbit.

Every cycle has advantages and disadvantages, and choosing the best one requires weighing considerations including cost, dependability, and efficiency. Before delving into the more sophisticated pump-fed cycles that power the rockets we see launching significant space missions today, let's take a quick look at cold gas and pressure-fed engines.

A Synopsis of Cold Gas and Pressure-Fed Engines

Let's examine these basic cycles before moving on to pump-fed engines. Pump-fed engines offer reliable performance with fewer parts, but they have limited power and efficiency.

The most basic rocket engines are cold gas thrusters. Cold gas thrusters store pressurized gas and discharge it through a nozzle to provide thrust. They are dependable and safe because there is no combustion involved. The temperature decreases when gasses expand. This phenomenon is known as the **Joule-Thomson effect**. Minor spacecraft changes, such as satellite attitude control, frequently employ cold gas thrusters. However, in comparison to other cycle types, they provide a tiny amount of thrust. This is used by NASA's Manned Maneuvering

Unit (MMU).

High-pressure gas, usually helium, forces the propellant from the tank into the combustion chamber of **pressure-fed engines**. Because it doesn't require pumps, this system is comparatively simple and dependable, making it appropriate for upper stages or smaller rockets. However, tank pressure constraints pressure-fed systems, preventing them from providing the high thrust levels needed for larger rockets. Additionally, the structural weight of the tank, which must support high pressure, constrains them.

These cycles aren't appropriate for rockets that are flying long distances or carrying huge payloads, but they do well for certain low-thrust applications. Pump-fed cycles are what we use for high-power applications.

Pump-Fed Cycle Engines

Pump-fed cycle engines, which use pumps to transfer fuel and oxidizer into the combustion chamber under high pressure, enable higher thrust and more efficient combustion. Efficiency increases with the amount of heat and pressure in the combustion chamber. Since these engines form the core of heavy-lift rockets, they are perfect for missions that demand a lot of power. Let's dissect the primary kinds of pump-fed cycles and examine their distinctive features.

Open Cycle/Gas Generator

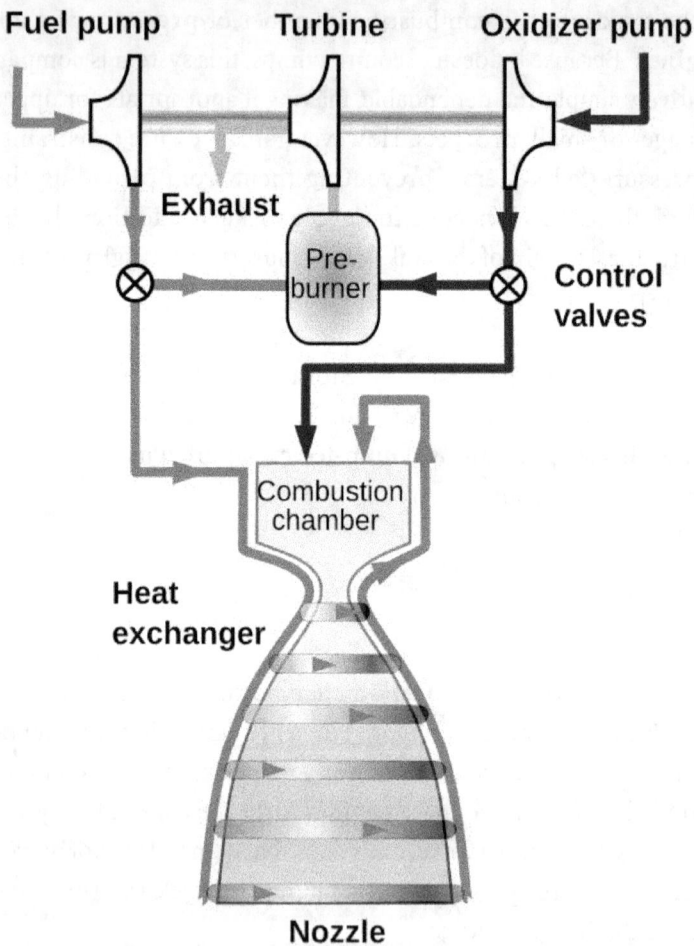

Source: https://commons.wikimedia.org/wiki/File:Gas_generator_rocket_cycle.svg

In an open cycle, also known as a gas generator cycle, a different, smaller chamber burns part of the propellant. The exhaust from this process then exposes the turbine that powers the pump. Instead of entering the combustion chamber, the exhaust is "dumped," or discharged. The German V-2's A4 engine is the first of this type. Types of propellants and examples: People frequently use RP-1, a type of kerosene, and liquid oxygen (LOX) for open cycles. The Falcon 9 rocket's SpaceX Merlin engines are well-known instances of open-cycle engines.

Benefits: Open cycles are robust and somewhat easy. They cost less than other cycles because they're easy to develop and produce. Additionally, by evacuating part of the gas instead of injecting it into the primary combustion chamber, it avoids some of the severe pressures and temperatures found in closed cycles.

Limitations: Open cycles often exhibit lower efficiency due to the gas generator's expenditure of some propellant. Although their reduced efficiency makes them less suitable for long-distance, high-performance missions, their dependability keeps them in widespread usage. Notable examples of open-cycle engines are the RD-107A & RD-108A engines on the Soyuz, the RD-68 engine on the Delta IV Heavy, the Merlin 1D engine on the Falcon 9, and the F-1 & J-2 engines on the Saturn V.

Closed-Cycle / Stage Combustion

Reusing all of the combustion gasses and sending them into the main combustion chamber increases efficiency in a closed cycle, also known as a staged combustion cycle. High-performance rockets use closed cycles, which are more intricate.

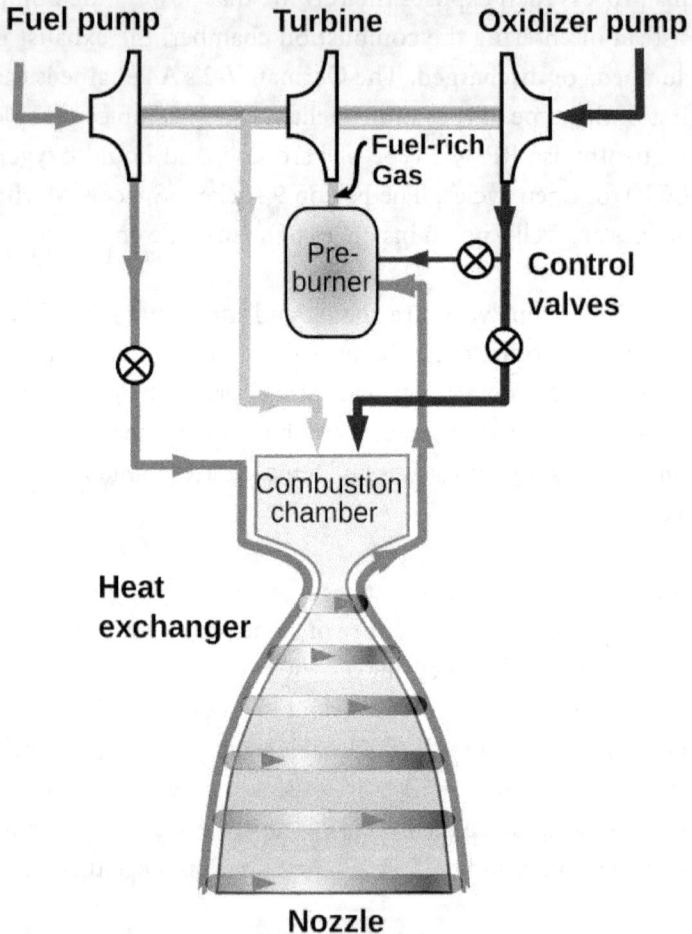

Source: https://commons.wikimedia.org/wiki/File:Staged_combustion_rocket_cycle.svg

It is difficult to direct the exhaust into the primary combustion chamber. Compared to the primary combustion chamber, the pressure downstream of the exhaust pipe would be lower. Gasses from the combustion chamber would backflow up the exhaust pipe as a result. If the fuel contains long-chain hydrocarbons like RP-1, the exhaust from the gas generator contains enough soot to clog the injectors and damage the engine.

The engine will pump the oxidizer or the fuel completely via the gas generator and the turbine to avoid soot damage, rather than pumping the exhaust to the primary combustion chamber. A tiny quantity of fuel flow through the turbine will cause the gas generator to react. We will now refer to the gas generator as a pre-burner. The type of closed-cycle fuel-rich or oxidizer-rich system will rely on the propellant flow through the pre-burner.

Rich-Oxygen Staged Combustion

This cycle produces a hot, oxygen-rich gas that powers the pumps before being pumped into the main combustion chamber by burning a tiny amount of fuel with an excess of oxygen in the pre-burner. The USSR overcame the difficulties of this cycle in the early 1950s. The USSR achieved this by allowing the pre-burner in the S1.5400, R7 upper-stage engine to activate with a required minimal amount of fuel. For instance, the Atlas V's Russian rocket engine, the RD-180, is a staged combustion engine that is rich in oxygen.

Pre-Burner Pressure

To guarantee that the gasses enter the main combustion chamber smoothly, the pre-burner keeps the pressure higher than the main chamber. Stable and effective combustion requires a controlled flow of high pressure from a region of lower pressure. Generally, the pre-burner pressure should be twice that of the main chamber, and the pressure at the back of the injector should be 20% higher than that of the main combustion chamber. This cycle compresses the oxidizer to high pressure to reach this pre-burner pressure level. Certain engines use several pump stages to manage high pressures and flow rates. Although they increase complexity and weight, multi-stage pumps provide more control over fuel flow and combustion stability.

Benefits include high efficiency and minimal propellant waste. This cycle does a great job of protecting the engine from soot and coking. **Limitations**: This cycle is difficult to create and maintain since it produces exhaust fumes that are highly hot and corrosive, which can be difficult for materials to stand with. When this highly heated oxygen is present, it reacts with everything, causing damage to the environment.

Fuel-Rich Staged Combustion Cycle

This cycle considers the alternative to what we previously examined. Instead of using oxygen in the pre-burner, this variation uses an excess of fuel. Using long-chain hydrocarbons like RP-1 leads to coking and soot accumulation, but substituting with non-carbon-rich fuel makes the cycle more reliable. Liquid hydrogen, being a light molecule without carbon, solves the problem. A fuel-rich staged combustion cycle powers the SpaceX Raptor engine, which powers the Starship rocket. This cycle reduces engine part wear by producing cooler exhaust gasses. The fuel-rich cycle requires complex balancing to ensure the right mix of fuel and oxidizer in the main combustion chamber.

Dual Pre-Burners

To maximize the balance of gasses entering the combustion chamber, certain staged combustion engines employ two pre-burners, one rich in fuel and the other rich in oxygen. Although dual pre-burners are more efficient, they are more complicated and have a higher risk of mechanical failure. The RS-25 engine is a wonderful example. Having high-pressure liquid hydrogen and liquid oxygen is a disaster waiting to happen.

Purge Seals

Purge seals stop oxidizer and fuel from seeping into inappropriate places. Purge seals aid in maintaining exact control over gas flow in fuel-rich staged combustion engines, which is essential for averting failures or explosions. The N1 rocket's RD-56 and RD-57 (USSR), the RD-0120 (USSR), and the RS-25 (USA) are all excellent illustrations of the fuel-rich staged combustion cycle.

Full-Flow Staged Combustion Cycle

This cycle maximizes power and economy by completely burning the fuel and oxidizer in separate pre-burners. The pre-burner and the turbine receive all of the fuel and oxidizer.

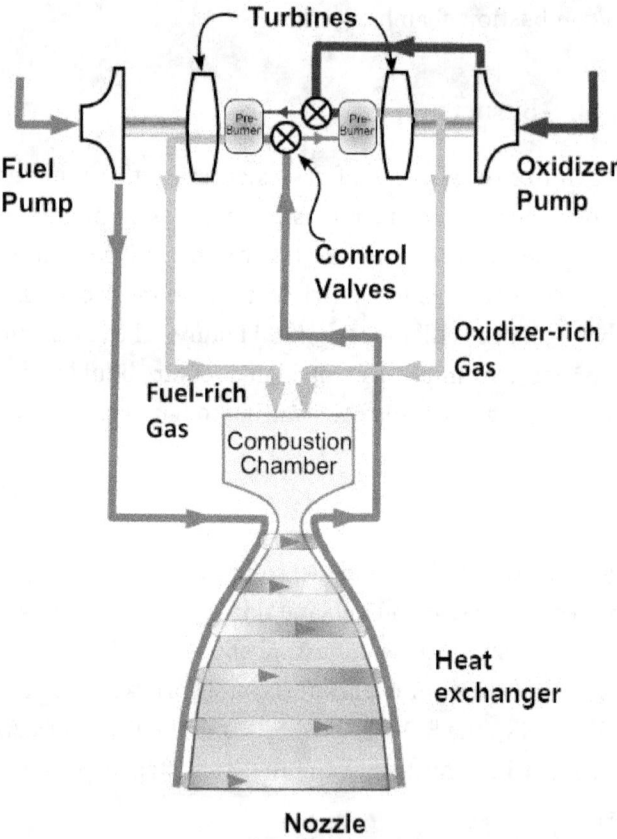

Source: https://en.wikipedia.org/wiki/File:Full_flow_staged_rocket_cycle.png

The structure design incorporates both fuel-rich and oxygen-rich pre-burners. Both oxygen and fuel enter the main chamber as gas, which lowers the residue and improves combustion efficiency. Gas plus gas is much easier to mix than liquid plus liquid or liquid plus gas.

The SpaceX Raptor employs a full-flow staged combustion cycle, making it unique among engines. The SpaceX Raptor is a worthy example. A pump-fed cycle achieves maximum thrust and efficiency. Full-flow cycles are costly to design and difficult to maintain due to their enormous complexity and need for sophisticated materials and engineering.

Pairing the fuel-rich turbine and shaft with the fuel pump eliminates the seal mechanism, as does pairing the oxidizer-rich turbine and shaft with the oxidizer pump. Since the full-flow and other close cycles operate under the same level of enthalpy to power the pump, the full-flow cycle allows more mass to flow through the burner and turbine, which is an ultimate advantage. The only engines with full-flow cycles are the SpaceX Raptor and the RD-270 (USSR).

The Expander Cycle

The heat from the combustion chamber's walls transforms the fuel into gas in an expander cycle, powering the pumps. The fuel absorbs heat as it circulates through the combustion chamber, eventually vaporizing. Before entering the combustion chamber, vaporized fuel powers the pumps. Expander cycles are more dependable than staged combustion engines because they are easier to understand and cooler.

Because they depend on the heat absorbed from the chamber walls, which limits scalability, they are only suitable for engines with lesser thrust. They are mostly found in smaller rockets, such as the RL10 engine in the Centaur upper stage. The Japanese H-I and HII rockets, the LE-5A and LE-5B, and Blue Origin's BE-3U (New Glenn upper stage)

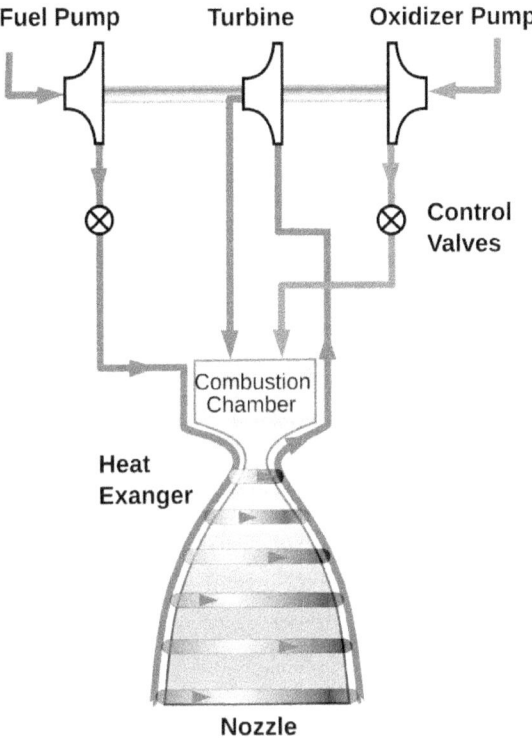

Source: https://en.wikipedia.org/wiki/Expander_cycle#/media/File:Expander_rocket_cycle.svg

Tap-Off Cycle

During the tap-off cycle, we divert part of the combustion gas to operate the pumps. This engine lowers the overall weight by eliminating the gas generator or pre-burner. Because this engine has no moving parts, it uses fuel flowing through the walls to regeneratively cool itself. As a result, the temperature of the main combustion chamber can reach 3500K.

Running an engine can be extremely hot. You can usually control this by adding gasoline-like fuel, which lowers the temperature and releases fuel-rich exhaust. It reduces the weight of the engine and is relatively simple to implement. However, it is less effective than staged combustion cycles and generates less thrust, which makes it less suitable for heavy-lift applications. The best examples are Blue Origin's BE-3 and NASA's J-2 (Saturn V 2nd stage).

ENGINES CYCLES

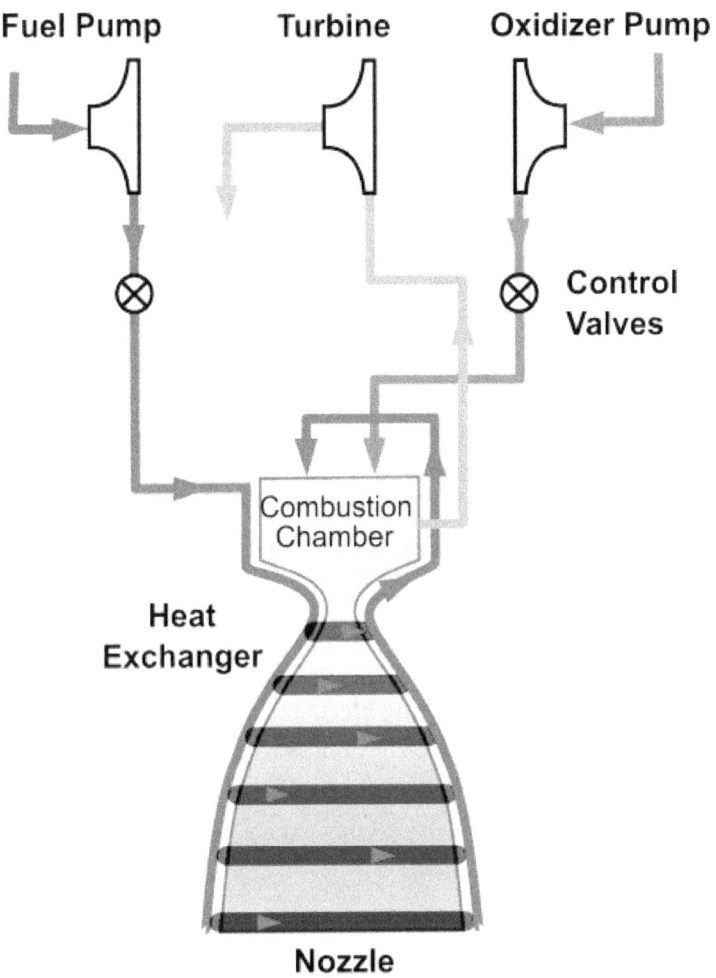

Source: https://commons.wikimedia.org/wiki/File:Combustion_tap-off_rocket_cycle.svg

Electric Pump-Fed Cycle

Electric motors, not gas, drive the pumps in electric pump-fed cycles. We can achieve higher pressure without raising the propellant tank pressure. This is where the electric pump comes into play.

An electric motor powers the pumps, eliminating the need for pre-burners and streamlining the system's high electricity consumption, typically from batteries. Battery weight may therefore be a limiting factor. Electric cycles are suitable for small rockets due to their ease of on and off operation.

They are unable to match the high power of conventional pump-fed cycles due to battery capacity limitations. The pumps on an RD-170 need 230,000 horsepower. An electric pump cannot produce this amount of power. These pumps are mostly found in tiny launchers, such as the Electron rocket from Rocket Lab. Rocket Lab has used electric pumps in electron rockets, and Astra's Delphin engines serve as excellent examples of this pump cycle.

In brief

Modern rocketry relies heavily on pump-fed cycles. Pump-fed cycles are essential for missions requiring great power and economy, even though simpler cycles like cold gas thrusters and pressure-fed engines have specialized uses. Here's a summary:

Open Cycle: Easy and dependable, but waste gas reduces efficiency.

Staged combustion, or closed cycle, is an efficient process that comes in a variety of forms, such as fuel-rich and oxygen-rich.

The Expander Cycle, which offers cooling and reliability, is only suitable for smaller engines.

The Tap-Off Cycle is simple and lightweight, but it is less effective.

The electric pump-fed cycle is very easy to use, but it is limited by battery power.

Every cycle has special benefits that meet certain mission requirements. We can better comprehend the trade-offs engineers make to push the boundaries of space exploration.

4

Real-World Applications

After discussing the basic principles of rocket engine cycles, it's time to observe how they function in real-world scenarios. From missiles from the Cold War to the state-of-the-art rockets being constructed today for Mars missions, these distinct cycles have been employed in a variety of ways over the years. This chapter explores the historical and current applications of each type of engine cycle in rocketry, along with potential future paths.

Early Day (Cold War Era)

Rocket technology rapidly improved during the Cold War era as the US and the USSR competed to create cutting-edge rockets for space exploration and military applications. During this period, the primary focus of rocket engine development was on long-distance payload (and occasionally warhead) delivery, leading to the development of early space flights and intercontinental ballistic missiles (ICBMs).

Simpler open-cycle and pressure-fed engines were prevalent in the early days of rocketry. Although not as effective as contemporary cycles, these designs were simple, reliable, and straightforward to implement. For missiles like the V-2 Rocket, which Germany developed during World War II and set the stage for later rocket designs, pressure-fed systems in particular were perfect since they were simple to design and maintain.

To increase rocket power and dependability, researchers started experimenting with more effective cycles, including staged combustion, in the late 1950s and early 1960s. With its oxygen-rich staged combustion engines and strong closed-cycle architecture, the Soviet Union produced the RD-170 series, which was incredibly efficient. The Soviets were able to reach remarkable payload capacities thanks to these engines, which served as the basis for launch vehicles such as the Proton rocket, which backed Soviet space operations for many years.

The race to the moon sped up the development of rocket engines in the United States. In its multi-stage design, NASA's Saturn V rocket, which carried Apollo missions, featured both open-cycle and closed-cycle engines. The J-2 engines (expander cycle) fueled the Saturn V's second stage, propelling the Apollo astronauts into lunar orbit and setting a new standard for reusable upper-stage engines. As each superpower fought for supremacy in space and military technologies, the Cold War era led to notable breakthroughs in engine cycles. Many of these advancements made possible the more advanced engines we use today.

Contemporary Uses

Rocket engines now are very different from what they were during the Cold War. Contemporary rockets are more dependable, efficient, and able to carry out challenging missions. The commercial and governmental sectors are increasingly using pump-fed cycles, such as full-flow staged combustion and staged combustion.

With the use of a full-flow staged combustion cycle in its Raptor engine, SpaceX has completely transformed the commercial space sector. The Starship rocket, built for big payload missions and Mars exploration, is powered by this engine. Full-flow staged combustion is the perfect option for SpaceX's ambitious aspirations because it is very efficient and provides excellent thrust with little waste. Setting a new benchmark in rocketry, the Raptor engine's two pre-burners maximize performance and reusability by burning both fuel-rich and oxygen-rich mixes.

Blue Origin, another contemporary participant, created the BE-4 engine for their New Glenn rocket. The BE-4 prioritizes dependability and reusability through the use of a staged combustion cycle. Due to its outstanding efficiency and endurance, the closed-cycle design supports Blue Origin's mission to lower launch costs and increase the sustainability of space access.

For its missions, NASA still uses closed-cycle engines and sophisticated expanders. For example, the RL10 engine, an expander cycle engine renowned for its dependability and efficiency, powers the Centaur upper stage. Because of its capacity to give precise control over vast distances, the RL10 has powered numerous space missions, including satellites and interplanetary probes.

Smaller rockets, like Rocket Lab's, have adopted the electric pump-fed cycle to simplify operations. By using battery-operated electric pumps, Rocket Lab's Electron rocket does away with complicated pre-burners and reduces engine weight. Small satellites, where flexibility is crucial but payload capacity is limited, benefit most from this method.

Contemporary applications display a wide variety of cycles, each selected for a particular objective. Today's rockets are expanding the realm of what we previously believed was feasible, whether it be for commercial satellite launches, deep space exploration, or crewed missions.

Upcoming Patterns

Future engine cycles will witness even more fascinating advancements. The need for effective, potent, and reusable engines will only increase as we pursue interplanetary travel and the possibility of establishing human settlements on other worlds. The following significant trends could influence rocket engine cycles in the future:

The future of rocketry is reusability. SpaceX and Blue Origin are pushing the boundaries by developing rockets that can launch, land, and then launch again. Future engines must be able to endure several cycles of high heat and cooling without suffering appreciable damage. This drive for reusability may revolve around full-flow staged combustion engines, which are extremely efficient and lessen the strain on engine parts.

Engines that can transition between several cycle modes may become useful as missions get more complicated. For instance, hybrid engines might begin at lower altitudes in an open-cycle mode and switch to a more efficient closed-cycle mode at higher altitudes. Rockets could operate at peak efficiency in a variety of mission scenarios because of these versatile engines.

Electric pump-fed cycles might become increasingly popular as battery technology advances, particularly for compact launch vehicles. Future designs may explore green propellant fuels, which emit fewer pollutants and are safer to handle. As environmental effects and space debris worries increase, "green" propulsion may become the norm.

Nuclear thermal propulsion engines may be a key component of deep-space exploration. These engines have exceptionally high efficiency for long-distance travel to worlds like Mars or even outside of our solar system because they superheat a propellant using nuclear processes. Although nuclear propulsion is still in the experimental stage, it is an important field of study since it could lead to faster and more energy-efficient travel.

Government agencies and commercial businesses will need engine cycles that can function in Martian conditions and generate sufficient thrust for return trips as they pursue Mars. SpaceX's Starship's Raptor engines, designed to function both on Mars and in Earth's atmosphere, serve as a prime example.

In order to make round-trip missions to Mars more practical, future engines may even use Martian resources as propellants, a concept known as In-Situ Resource Utilization, or ISRU.

Rocketry has a promising future because of the quick developments in fuel and engine design, which open up new possibilities. To propel humanity farther into space than ever before, cycles that strike a balance between effectiveness, reusability, and adaptability will be crucial.

This chapter discussed the practical uses of rocket engine cycles, demonstrating how earlier inventions impact current designs and how potential future technological advancements could increase our reach in space. The motors driving humanity's ascent to the heavens are the engine cycles we've covered, and they go beyond simple technological ideas.

5

Conclusion

You deserve congratulations for reaching the last chapter! You now have a firm understanding of the fundamentals of rocket engine cycles. You now have a better understanding of the fundamentals behind contemporary rocketry, from the parts and propellants that power these strong machines to the engine cycles that optimize their thrust and efficiency.

This concluding chapter will provide a summary of the key points, a look into engine cycles' future, and an analysis of why it is essential to comprehend these ideas, even if you are new to the subject. We have explored the fundamentals of rocket engine cycles in this book in an approachable manner. Let's go over the main points we discussed:

We started with a quick overview of rocketry's history, including everything from the first fireworks rockets to the Cold War space competition. We then presented the action-reaction principle, a straightforward yet effective idea that explains how rockets lift off. We examined the main parts of a rocket engine, including the combustion chamber, pump, propellant tank, and nozzle, and how each contributes differently to thrust production. We then discussed the types and characteristics of solid, liquid, and hybrid propellants, highlighting the significant influence propellant selection has on engine design overall.

We delved deeply into engine cycles in Chapter 3, concentrating mostly on pump-fed engines. We talked about tap-off, expander, electric pump-fed, open-cycle, and closed-cycle (staged combustion) cycles. We described the benefits, drawbacks, and practical uses of each cycle.

We looked at the historical uses of various engine cycles, their present uses, and the exciting trends influencing the future. We discussed the innovative ways in which organizations such as SpaceX, Blue Origin, Rocket Lab, and NASA are developing rockets that are more potent, effective, and reusable than in the past.

We've illustrated in these sections how every engine cycle decision affects the rockets' total performance, efficiency, and mission capabilities. Rocket engine cycles vary widely, each suited to the specific requirements of the mission, from the tiny cold gas thrusters on satellites to the potent Raptor engines made for interplanetary travel.

Rocket Engine Cycles' Future

The technology for rocket engines is developing quickly, and there are a lot of exciting things to come. We may anticipate seeing more engines made with longevity and reusability in mind as reusable rockets become the norm in the industry. This push for reusability is likely to encourage more full-flow staged combustion cycles and hybrid engines designed to withstand numerous launches and landings.

Green propellants and electric pump-fed cycles might become more popular as people's interest in environmentally friendly alternatives grows, particularly for smaller launch vehicles. In the long run, space access will be more sustainable thanks to green propulsion technology, which promises to lessen the environmental impact of launches.

Rocket engines must be able to withstand lengthy flights and possibly make use of local resources for human trips to Mars and beyond. Technologies such as ISRU (In-Situ Resource Utilization) and nuclear thermal propulsion are likely to enable interplanetary travel. Adaptable cycles in future engines could enable rockets to change modes according to mission stage or altitude. This adaptability might increase productivity and create new opportunities for intricate, multi-phase missions.

Rocketry has a bright future, and new designs and engine cycles that challenge our existing understanding are certain to emerge as developments continue. While comprehending the cycles of today is only the first step, it offers a solid basis for understanding our future.

Why concentrate on engine cycles, you ask? For both novices and enthusiasts, comprehending these cycles involves more than simply mastering technical terms; it also entails understanding the design concepts that enable space travel. Built to

CONCLUSION

overcome the enormous problems of gravity, fuel efficiency, and the vacuum of space, each rocket engine cycle is a testimony to human ingenuity.

Understanding engine cycles illustrates the amazing planning, experimentation, and precision that go into every launch, regardless of your interest in science, engineering, or rocketry. Every cycle, part, and design decision pushes the limits of human potential and brings us one step closer to the stars.

This information might even encourage you to learn more about rocketry, work in the sector, or help shape the next wave of aerospace technological advancements. The possibilities are endless.

I appreciate you coming along with me as we explore the intriguing realm of rocket engine cycles. I hope this book has provided you with a welcoming introduction to a subject that can appear daunting at first glance. Please think about posting a review on Amazon if you have liked learning about these ideas; it will help other readers find this book and will also let me know whether the content was enjoyable and useful.

Above all, keep in mind that the journey is far from over. Even though rocket science can be difficult, anyone can learn more about the mechanisms that propel our exploration of the universe if they are curious and eager to study.

Continue to look up, and raise a toast to the adventure beyond Earth!

6

Reference

1. A review of United States Air Force and Department of Defense aerospace propulsion needs. (2006). *National Academies Press eBooks*. https://doi.org/10.17226/11780
2. dev@torro.io. (2024, October 6). A comparison of different rocket engine cycles throughout the years | SoftInWay. *SoftInWay*. https://www.softinway.com/a-comparison-of-different-rocket-engine-cycles-throughout-the-years
3. *Gas-Generator Cycle | Space4Water Portal*. (n.d.). https://space4water.org/space/gas-generator-cycle
4. Law, A., & Kordina, F. (2023, February 17). *Rocket engine cycles*. Everyday Astronaut. https://everydayastronaut.com/rocket-engine-cycles/
5. Leishman, J. G. (2023, January 1). *Rocket engines*. Pressbooks. https://eaglepubs.erau.edu/introductiontoaerospaceflightvehicles/chapter/rocket-engines/
6. NASA Glenn Research Center. (2023, April 17). *Chemical

Propulsion Systems | Glenn Research Center | NASA. Glenn Research Center | NASA. https://www1.grc.nasa.gov/research-and-engineering/chemical-propulsion-systems/
7. Price, E. W., & Biblarz, O. (2024, November 5). *Rocket | Characteristics, propulsion, development, & facts*. Encyclopedia Britannica. https://www.britannica.com/technology/rocket-jet-propulsion-device-and-vehicle/Liquid-propellant-rocket-engines
8. *Rocket Propulsion*. (n.d.). https://www.grc.nasa.gov/www/k-12/airplane/rocket.html
9. Tinelmis. (2024, April 7). Gas-generator Cycle - Definition & Detailed Explanation - Rocketry & Propulsion Glossary - Sentinel. *Sentinel Mission*. https://sentinelmission.org/rocketry-propulsion-glossary/gas-generator-cycle/
10. Wikipedia contributors. (2024, September 26). *Liquid-propellant rocket*. Wikipedia. https://en.wikipedia.org/wiki/Liquid-propellant_rocket

www.ingramcontent.com/pod-product-compliance
Lightning Source LLC
Chambersburg PA
CBHW050317220526
45465CB00005B/2030